Big Grassland Animals

Katie Peters

GRL Consultants,
Diane Craig and Monica Marx,
Certified Literacy Specialists

Lerner Publications ◆ Minneapolis

Lerner Publications Company
A division of Lerner Publishing Group, Inc.
241 First Avenue North
Minneapolis, MN 55401 USA

For reading levels and more information, look up this title at www.lernerbooks.com.

Main body text set in Memphis Pro 24/39
Typeface provided by Linotype.

Photo Acknowledgments
The images in this book are used with the permission of: © iStockphoto (all)

Front cover: © Shutterstock

Library of Congress Cataloging-in-Publication Data

Names: Peters, Katie, author.
Title: Big grassland animals / Katie Peters.
Description: Minneapolis : Lerner Publications, [2020] | Series: Let's look at animal
    habitats (Pull ahead readers - Nonfiction) | Includes index. | Audience: Age 4–7. |
    Audience: K to Grade 3.
Identifiers: LCCN 2018059820 (print) | LCCN 2019000027 (ebook) | ISBN 9781541562028
    (eb pdf) | ISBN 9781541558625 (lb : alk. paper) | ISBN 9781541573116 (pb : alk. paper)
Subjects: LCSH: Grassland animals—Juvenile literature.
Classification: LCC QL115.3 (ebook) | LCC QL115.3 .P48 2020 (print) | DDC 591.74—dc23

LC record available at https://lccn.loc.gov/2018059820

Manufactured in the United States of America
1 – CG – 7/15/19

Contents

Big Grassland Animals

Here is a big elephant.

Here is a big lion.

Here is a big giraffe.

Here is a big zebra.

Here is a big kangaroo.

Here is a big bison.

Did You See It?

bison

elephant

giraffe

kangaroo

lion

zebra

Index